Made in the United States
Text printed on 100%
recycled paper

Houghton
Mifflin
Harcourt

Printed in the U.S.A.

ISBN 978-0-544-29551-3

18   0877   20

4500800212   B C D E F G

Dear Students and Families,

Welcome to **Go Math!**, Grade 6! In this exciting mathematics program, there are hands-on activities to do and real-world problems to solve. Best of all, you will write your ideas and answers right in your book. In **Go Math!**, writing and drawing on the pages helps you think deeply about what you are learning, and you will really understand math!

By the way, all of the pages in your **Go Math!** book are made using recycled paper. We wanted you to know that you can Go Green with **Go Math!**

Sincerely,

The Authors

Made in the United States
Text printed on 100% recycled paper

Authors

Juli K. Dixon, Ph.D.
Professor, Mathematics Education
University of Central Florida
Orlando, Florida

Edward B. Burger, Ph.D.
President, Southwestern University
Georgetown, Texas

Steven J. Leinwand
Principal Research Analyst
American Institutes for
    Research (AIR)
Washington, D.C.

Contributor

Rena Petrello
Professor, Mathematics
Moorpark College
Moorpark, California

Matthew R. Larson, Ph.D.
K-12 Curriculum Specialist for
    Mathematics
Lincoln Public Schools
Lincoln, Nebraska

Martha E. Sandoval-Martinez
Math Instructor
El Camino College
Torrance, California

English Language Learners Consultant

Elizabeth Jiménez
CEO, GEMAS Consulting
Professional Expert on English
    Learner Education
Bilingual Education and
    Dual Language
Pomona, California

The Number System

 Critical Area Completing understanding of division of fractions and extending the notion of number to the system of rational numbers, which includes negative numbers

1 Whole Numbers and Decimals   3

COMMON CORE STATE STANDARDS

6.NS The Number System
Cluster B Compute fluently with multi-digit numbers and find common factors and multiples.
6.NS.B.2, 6.NS.B.3, 6.NS.B.4

GO DIGITAL

Go online! Your math lessons are interactive. Use *i*Tools, Animated Math Models, the Multimedia eGlossary, and more.

Essential Question
How do you divide multi-digit numbers?
Start

Chapter 1 Overview

In this chapter, you will explore and discover answers to the following **Essential Questions**:

• How do you solve real-world problems involving whole numbers and decimals?

• How does estimation help you solve problems involving decimals and whole numbers?

• How can you use the GCF and the LCM to solve problems?

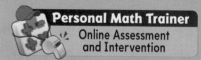

Personal Math Trainer
Online Assessment and Intervention

FOR MORE PRACTICE
GO TO THE
Personal Math Trainer

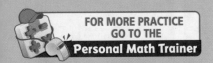

Practice and Homework

Lesson Check and
Spiral Review in
every lesson

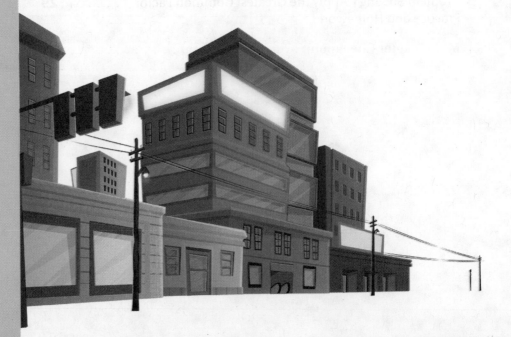

The Number System

Common Core

CRITICAL AREA Completing understanding of division of fractions and extending the notion of number to the system of rational numbers, which includes negative numbers

Sweet Success

Businesses that sell food products need to combine ingredients in the correct amounts. They also need to determine what price to charge for the products they sell.

Get Started

A company sells Apple Cherry Mix. They make large batches of the mix that can be used to fill 250 bags each. Determine how many pounds of each ingredient should be used to make one batch of Apple Cherry Mix. Then decide how much the company should charge for each bag of Apple Cherry Mix, and explain how you made your decision.

Important Facts

Ingredients in Apple Cherry Mix (1 bag)
- $\frac{3}{4}$ pound of dried apples
- $\frac{1}{2}$ pound of dried cherries
- $\frac{1}{4}$ pound of walnuts

Cost of Ingredients
- dried apples: $2.80 per pound
- dried cherries: $4.48 per pound
- walnuts: $3.96 per pound

Completed by _____

Whole Numbers and Decimals

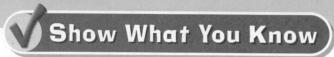

✓ Show What You Know

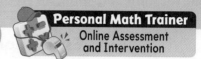

Personal Math Trainer
Online Assessment
and Intervention

Check your understanding of important skills.

Name _____

▶ **Factors** **Find all of the factors of the number.** (4.OA.B.4)

1. 16 _____

2. 27 _____

3. 30 _____

4. 45 _____

▶ **Round Decimals** **Round to the place of the underlined digit.** (5.NBT.A.4)

5. 0.$\underline{3}$23

6. $\underline{4}$.096

7. 1$\underline{0}$.67

8. 5.2$\underline{7}$8

▶ **Multiply 3-Digit and 4-Digit Numbers** **Multiply.** (5.NBT.B.5)

9. 2,143
 $\times$ 6

10. 375
 $\times$ 8

11. 3,762
 $\times$ 7

12. 603
 $\times$ 9

Math in the Real World

Maxwell saved $18 to buy a fingerprinting kit that costs $99. He spent 0.25 of his savings to buy a magnifying glass. Help Maxwell find out how much more he needs to save to buy the fingerprinting kit.

▶ **Visualize It** ••

Complete the Flow Map using the words with a ✓.

Estimation

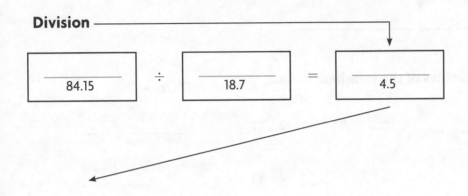

Division

| 84.15 | ÷ | 18.7 | = | 4.5 |

| 80 | ÷ | 20 | = | 4 |

▶ **Understand Vocabulary** ••••••••••••••••••••••••••••••

Complete the sentences using the preview words.

1. The least number that is a common multiple of two or more

   numbers is the _____.

2. The greatest factor that two or more numbers have in common

   is the _____.

3. A number that is a factor of two or more numbers is a

   _____.

4. A number written as the product of its prime factors is the

   _____ of the number.

GO DIGITAL
• **Interactive Student Edition**
• **Multimedia eGlossary**

Chapter 1 Vocabulary

common factor

factor común

12

dividend

dividendo

23

divisor

divisor

25

factor

factor

33

greatest common factor (*GCF*)

máximo común divisor (MCD)

38

least common multiple (*LCM*)

mínimo común múltiplo (m.c.m.)

48

prime factorization

descomposición en factores primos

77

prime number

número primo

78

The number that is to be divided in a division problem

$18 \div 6$ 6⟌18

↑   ↑
dividend   dividend

A number that is a factor of two or more numbers

Example:

Factors of 16: 1, 2, 4, 8, 16

Factors of 20: 1, 2, 4, 5, 10, 20

A number multiplied by another number to find a product

$2 \times 3 \times 7 = 42$

↑ ↑ ↑
factors

The number that divides the dividend

$18 \div 6$ 6⟌18

↑   ↑
divisor   divisor

The least number that is a common multiple of two or more numbers

Example:

Multiples of 4: 4, 8, 12, 16, 20, 24, 28, 32, 36

Multiples of 6: 6, 12, 18, 24, 30, 36, 42

The LCM of 4 and 6 is 12.

The greatest factor that two or more numbers have in common

Example:

Factors of 18: 1, 2, 3, 6, 9, 18

Factors of 30: 1, 2, 3, 5, 6, 10, 15, 30

The GCF of 18 and 30 is 6.

A number that has exactly two factors: 1 and itself

Examples: 2, 3, 5, 7, 11, 13, 17, and 19 are prime numbers. 1 is not a prime number.

A number written as the product of all its prime factors

Example:

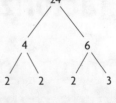

$24 = 2 \times 2 \times 2 \times 3$

Going to Washington, DC

For 2 to 4 players

Word Box
common factor
dividend
divisor
factor
greatest common factor (GCM)
least common multiple (LCM)
prime factorization
prime number

Materials

- 3 of 1 color per player: red, blue, green, and yellow playing pieces
- 1 number cube

How to Play

1. Put your 3 playing pieces in the START circle of the same color.
2. To get a playing piece out of START, you must toss a 6.
   - If you toss a 6, move 1 of your playing pieces to the same-colored circle on the path. If you do not toss a 6, wait until your next turn.
3. Once you have a playing piece on the path, toss the number cube to take a turn. Move the playing piece that many tan spaces. You must get all three of your playing pieces on the path.
4. If you land on a space with a question, answer it. If you are correct, move ahead 1 space.
5. To reach FINISH move your playing pieces up the path that is the same color as your playing pieces. The first player to get all three playing pieces on FINISH wins.

START

How can you use a list to find the least common multiple of two numbers?

Explain how to use prime factorization to find a least common multiple.

FINISH

Name two common factors for 32 and 40.

What is the greatest common factor of 28 and 49?

How many factors does a prime number have?

Explain two ways to find the prime factorization of 210.

True or false:
The LCM of 7 and 14 is 42.
Explain how you know.

How can you find the common factors of two numbers?

START

Game

START

START

FINISH

Explain why a prime factorization may be used as a secret code.

2 × 2 × 3 × 5 is the prime factorization of what number?

What is the least common multiple of 4 and 10?

What is the dividend in 48 ÷ 8?

What is a least common multiple?

What is a greatest common factor?

Why might you estimate when a dividend has a decimal?

How can you change a divisor such as 16.3 into a whole number?

© Houghton Mifflin Harcourt Publishing Company   Image Credits: (bg) ©Image Jen/Corbis; (tl) ©tarajane/Fotolia; (bl) ©VanHart/Fotolia; (c), (br) ©PhotoDisc/Getty Images; (tr) ©GlowImages/Alamy

The Write Way

Reflect

Choose one idea. Write about it.

- Explain how to find the prime factorization of 140.
- Two students found the least common multiple of 6 and 16.
  Tell how you know which student's answer is correct.
  Student A: The LCM of 6 and 16 is 96.
  Student B: The LCM of 6 and 16 is 48.
- Describe two ways to find the greatest common factor of 36 and 42.
- Write and solve a word problem that uses a whole number as a
  divisor and a decimal as a dividend.

Name _____

Divide Multi-Digit Numbers

Essential Question How do you divide multi-digit numbers?

Common Core The Number System—
6.NS.B.2
MATHEMATICAL PRACTICES
MP1, MP2, MP6

🔑 Unlock the Problem

When you watch a cartoon, the frames of film seem to blend together to form a moving image. A cartoon lasting just 92 seconds requires 2,208 frames. How many frames do you see each second when you watch a cartoon?

 Divide $2,208 \div 92$.

Estimate using compatible numbers. _____ ÷ _____ = _____

$$\begin{array}{r} 2 \\ 92\overline{)2,208} \\ -1\,84\downarrow \\ \hline 368 \\ - \\ \hline \end{array}$$

Divide the tens.

Divide the ones.

Compare your estimate with the quotient. Since the estimate, _____,

is close to _____, the answer is reasonable.

So, you see _____ frames each second when you watch a cartoon.

🔑 Example 1 Divide $12,749 \div 18$.

Estimate using compatible numbers. _____ ÷ _____ = _____

STEP 1 Divide.

$$\begin{array}{r} 70\ \text{r}5 \\ 18\overline{)12,749} \\ -12\,6\downarrow \\ \hline 14 \\ -0\downarrow \\ \hline 149 \\ - \\ \hline \end{array}$$

STEP 2 Check your answer.

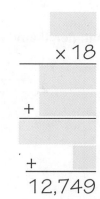

$$\times\ 18$$

$$+$$

$$+$$

$$12,749$$

Math Idea

You can write a remainder with an r, as a fractional part of the divisor, or as a decimal. For $131 \div 5$, the quotient can be written as 26 r1, $26\frac{1}{5}$, or 26.2.

Multiply the whole number part of the quotient by the divisor.

Add the remainder.

So, $12,749 \div 18 =$ _____.

🔑 Example 2

Divide 59,990 ÷ 280. Write the remainder as a fraction.

Estimate using compatible numbers. _____ ÷ _____ = _____

STEP 1 Divide.

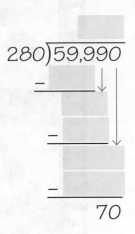

$$280\overline{)59{,}990}$$

70

STEP 2 Write the remainder as a fraction.

$$\frac{\text{remainder}}{\text{divisor}} = \frac{\boxed{}}{280}$$

Write the remainder over the divisor.

$$\frac{70 \div \boxed{}}{280 \div \boxed{}} = \frac{\boxed{}}{\boxed{}}$$

Simplify.

Compare your estimate with the quotient. Since the estimate, _____

is close to _____, the answer is reasonable.

So, 59,990 ÷ 280 = _____.

MATHEMATICAL PRACTICE ① **Describe** two ways to check your answer in Example 2.

_____

_____

_____

Share and Show

Estimate. Then find the quotient. Write the remainder, if any, with an r.

1. $29\overline{)986}$ Think: 30 × 3 = 90

2. $37\overline{)3{,}786}$

_____

_____

Name _____

Divide Multi-Digit Numbers

COMMON CORE STANDARD—6.NS.B.2
Compute fluently with multi-digit numbers and find common factors and multiples.

Estimate. Then find the quotient. Write the remainder, if any, with an r.

1.
$$\begin{array}{r} 13 \\ 55\overline{)715} \\ \underline{55} \\ 165 \\ \underline{165} \\ 0 \end{array}$$

Estimate:
$700 \div 50 = 15$

2. $19\overline{)800}$

3. $68\overline{)1,025}$

_____ _____

Estimate. Then find the quotient. Write the remainder, if any, as a fraction.

4. $20\overline{)1,683}$

5. $14,124 \div 44$

6. $11,629 \div 29$

_____ _____ _____

Find the least whole number that can replace ▮ to make the statement true.

7. $\blacksquare \div 7 > 800$

8. $\blacksquare \div 21 > 13$

9. $15 < \blacksquare \div 400$

_____ _____ _____

Problem Solving Real World

10. A plane flew a total of 2,220 miles. Its average speed was 555 miles per hour. How many hours did the plane fly?

11. A van is carrying 486 pounds. There are 27 boxes in the van. What is the average weight of each box in the van?

_____ _____

12. **WRITE** ▸*Math* Find $56,794 \div 338$. Write the quotient twice, once with the remainder as a fraction and once with an r.

_____

Lesson Check (6.NS.B.2)

1. A caterer's fee is based on the number of meals she provides. How much is the price per meal if the total fee is $1,088 for 64 meals?

2. Amelia needs 24 grams of beads to make a bracelet. She has 320 grams of beads. How many bracelets can she make?

Spiral Review (5.NBT.A.2, 5.NBT.A.3b, 5.NBT.B.7)

3. Hank bought 2.4 pounds of apples. Each pound cost $1.95. How much did Hank spend on the apples?

4. Gavin bought 4 packages of cheese. Each package weighed 1.08 kilograms. How many kilograms of cheese did Gavin buy?

5. Mr. Thompson received a water bill for $85.98. The bill covered three months of service. He used the same amount of water each month. How much does Mr. Thompson pay for water each month?

6. Layla used 0.482 gram of salt in her experiment. Maurice use 0.51 gram of salt. Who used the greater amount of salt?

FOR MORE PRACTICE
GO TO THE
Personal Math Trainer

Prime Factorization

Essential Question How do you write the prime factorization of a number?

**The Number System—
6.NS.B.4**

MATHEMATICAL PRACTICES
MP1, MP3, MP6, MP7

Unlock the Problem

Secret codes are often used to send information over the Internet. Many of these codes are based on very large numbers. For some codes, a computer must determine the prime factorization of these numbers to decode the information.

The **prime factorization** of a number is the number written as a product of all of its prime factors.

One Way Use a factor tree.

The key for a code is based on the prime factorization of 180. Find the prime factorization of 180.

Choose any two factors whose product is 180. Continue finding factors until only prime factors are left.

Remember

A prime number is a whole number greater than 1 that has exactly two factors: itself and 1.

A Use a basic fact.

Think: 10 times what number is equal to 180?

10 × _____ = 180

```
          180
         /    \
       10      ☐
      /  \    /  \
     2    ☐  6    ☐
    /    / \ / \   \
   2    ☐  ☐ 3  ☐
```

B Use a divisibility rule.

Think: 180 is even, so it is divisible by 2.

2 × _____ = 180

```
          180
         /    \
        2      ☐
       /  \   /  \
      2    2  ☐   3
     /    /  |  \  \
    ☐    ☐  ☐  ☐  ☐
```

180 = _____ × _____ × _____ × _____ × _____   List the prime factors from least to greatest.

So, the prime factorization of 180 is _____ × _____ × _____ × _____ × _____.

MATHEMATICAL PRACTICES **3**

Apply How can you apply divisibility rules to make the factor tree on the left?

🔒 Another Way Use a ladder diagram.

The key for a code is based on the prime factorization of 140. Find the prime factorization of 140.

Choose a prime factor of 140. Continue dividing by prime factors until the quotient is 1.

Ⓐ Use the divisibility rule for 2.

Think: 140 is even, so 140 is divisible by 2.

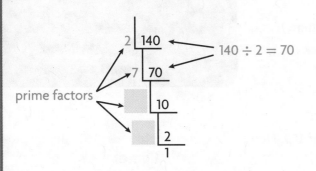

140 = _____ × _____ × _____ × _____

So, the prime factorization of 140 is _____ × _____ × _____ × _____.

Ⓑ Use the divisibility rule for 5.

Think: The last digit is 0, so 140 is divisible by 5.

```
5 | 140
  2 |___
     | 14
        | 2
        |___
```

List the prime factors from least to greatest.

Share and Show MATH BOARD

Find the prime factorization.

1. 18

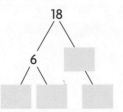

18 = _____ × _____ × _____

2. 42

```
2 | 42
  3 |___
     | 7
        |___
```

42 = _____ × _____ × _____

Math Talk

MATHEMATICAL PRACTICES ①

Describe a different strategy you could use to find the prime factorization of 140.

Find the prime factorization.

3. 75

✓ **4.** 12

✓ **5.** 65

_____

_____

_____

MATHEMATICAL PRACTICES ⑥

Explain why a prime number cannot be written as a product of prime factors.

On Your Own

Write the number whose prime factorization is given.

6. $2 \times 2 \times 2 \times 7$

7. $2 \times 2 \times 5 \times 5$

8. $2 \times 2 \times 2 \times 2 \times 3 \times 3$

_____

_____

_____

Practice: Copy and Solve **Find the prime factorization.**

9. 45

10. 50

11. 32

12. 76

13. 108

14. 126

15. The area of a rectangle is the product of its length and width. A rectangular poster has an area of 260 square inches. The width of the poster is greater than 10 inches and is a prime number. What is the width of the poster?

16. **MATHEMATICAL PRACTICE ⑦ Look for Structure** Dani says she is thinking of a secret number. As a clue, she says the number is the least whole number that has three different prime factors. What is Dani's secret number? What is its prime factorization?

Problem Solving • Applications Real World

Use the table for 17–19. Agent Sanchez must enter a code on a keypad to unlock the door to her office.

17. In August, the digits of the code number are the prime factors of 150. What is the code number for the office door in August?

_____

18. GO DEEPER In September, the fourth digit of the code number is 2 more than the fourth digit of the code number based on the prime factors of 225. The prime factors of what number were used for the code in September?

_____

19. THINK SMARTER One day in October, Agent Sanchez enters the code 3477. How do you know that this code is incorrect and will not open the door?

_____

Code Number Rules

Code Number Rules
1. The code is a 4-digit number.
2. Each digit is a prime number.
3. The prime numbers are entered from least to greatest.
4. The code number is changed at the beginning of each month.

WRITE Math • **Show Your Work**

20. THINK SMARTER Use the numbers to complete the factor tree. You may use a number more than once.

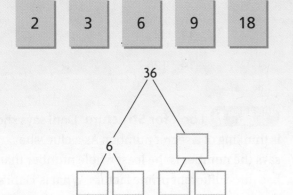

Write the prime factorization of 36.

_____

Prime Factorization

Common Core **COMMON CORE STANDARD—6.NS.B.4**
Compute fluently with multi-digit numbers and find common factors and multiples.

Find the prime factorization.

1. 44

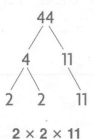

2 × 2 × 11

2. 90

_____

3. 48

_____

4. 204

_____

5. 400

_____

6. 112

_____

Problem Solving Real World

7. A computer code is based on the prime factorization of 160. Find the prime factorization of 160.

_____

8. The combination for a lock is a 3-digit number. The digits are the prime factors of 42 listed from least to greatest. What is the combination for the lock?

_____

9. WRITE ▸ *Math* Describe two methods for finding the prime factorization of a number.

_____

_____

_____

Lesson Check (6.NS.B.4)

1. Maritza remembers her PIN because it is between 1,000 and 1,500 and it is the product of two consecutive prime numbers. What is her PIN?

2. Brent knows that the 6-digit number he uses to open his computer is the prime factorization of 5005. If each digit of the code increases from left to right, what is his code?

Spiral Review (5.OA.A.2, 5.NBT.A.1, 5.NBT.B.6)

3. Piano lessons cost $15. What expressions could be used to find the cost in dollars of 5 lessons?

4. A jet plane costs an airline $69,500,000. What is the place value of the digit 5 in this number?

5. A museum has 13,486 butterflies, 1,856 ants, and 13,859 beetles. What is the order of the insects from least number to greatest number?

6. Juan is reading a 312-page book for school. He reads 12 pages each day. How long will it take him to finish the book?

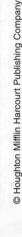

FOR MORE PRACTICE
GO TO THE
Personal Math Trainer

Least Common Multiple

Essential Question How can you find the least common multiple of two whole numbers?

 The Number System—
6.NS.B.4
MATHEMATICAL PRACTICES
MP2, MP3, MP4, MP6

 Unlock the Problem

In an experiment, each flowerpot will get one seed. If the flowerpots are in packages of 6 and the seeds are in packets of 8, what is the least number of plants that can be grown without any seeds or pots left over?

The **least common multiple**, or **LCM**, is the least number that is a common multiple of two or more numbers.

- Explain why you cannot buy the same number of packages of each item.

_____

_____

_____

One Way Use a list.

Make a list of the first eight nonzero multiples of 6 and 8. Circle the common multiples. Then find the least common multiple.

Multiples of 6: 6, 12, 18, _____,_____,_____,_____,_____

Multiples of 8: 8, 16, 24, _____,_____,_____,_____,_____

The least common multiple, or LCM, is _____.

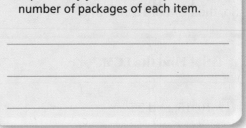

Another Way Use prime factorization and a Venn diagram.

Write the prime factorization of each number.

$6 = 2 \times$ _____

$8 = 2 \times$ _____ $\times$ _____

List the common prime factors of the numbers, if any.

Place the prime factors of the numbers in the appropriate parts of the Venn diagram.

To find the LCM, find the product of all of the prime factors in the Venn diagram.

$3 \times 2 \times 2 \times 2 =$ _____

The LCM is _____.

So, the least number of plants is _____.

6 and 8 have one prime factor of _____ in common.

Prime factors of 6   **Prime factors of 8**

3   2

Common prime factors

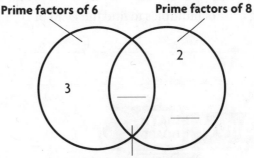

Math Talk   MATHEMATICAL PRACTICES ④

Model How does the diagram help you find the LCM of 6 and 8?

Example
Use prime factorization to find the LCM of 12 and 18.

Write the prime factorization of each number.

Line up the common factors.

Multiply one number from each column.

$12 = 2 \times 2 \times$ ____

$18 = 2 \quad \times \quad 3 \quad \times$ ____

$2 \times 2 \times \quad 3 \quad \times \quad 3 = 36$

Math Idea

The factors in the prime factorization of a number are usually listed in order from least to greatest.

So, the LCM of 12 and 18 is _____.

Try This! Find the LCM.

A 10, 15, and 25

Use prime factorization.

10 = _____

15 = _____

25 = _____

_____

The LCM is _____.

B 3 and 12

Use a list.

Multiples of 3: _____

Multiples of 12: _____

The LCM is _____.

---

1. How can you tell whether the LCM of a pair of numbers is one of the numbers? Give an example.

_____

_____

2. **MATHEMATICAL PRACTICE 6** **Explain** one reason why you might use prime factorization instead of making a list of multiples to find the LCM of 10, 15, and 25.

_____

_____

Share and Show

☑ 1. List the first six nonzero multiples of 6 and 9. Circle the common multiples. Then find the LCM.

Multiples of 6: _____

Multiples of 9: _____   The LCM of 6 and 9 is _____.

Name _____

Find the LCM.

2. 3, 5

3. 3, 9

✓ **4.** 9, 15

_____

_____

_____

On Your Own

Find the LCM.

Math Talk

MATHEMATICAL PRACTICES ⑥

Explain what the LCM of two numbers represents.

5. 5, 10

6. 3, 8

7. 9, 12

_____

_____

_____

MATHEMATICAL PRACTICE ② **Use Reasoning Algebra Write the unknown number for the ■.**

8. 5, 8   LCM: ■

9. ■, 6   LCM: 42

■ = _____

■ = _____

10. *THINK SMARTER* How can you tell when the LCM of two numbers will equal one of the numbers or equal the product of the numbers?

_____

_____

11. MATHEMATICAL PRACTICE ③ **Verify the Reasoning of Others** Mr. Haigwood is shopping for a school picnic. Veggie burgers come in packages of 15, and buns come in packages of 6. He wants to serve veggie burgers on buns and wants to have no items left over. Mr. Haigwood says that he will have to buy at least 90 of each item, since $6 \times 15 = 90$. Do you agree with his reasoning? Explain.

_____

_____

12. *GO DEEPER* A deli has a special one-day event to celebrate its anniversary. On the day of the event, every eighth customer receives a free drink. Every twelfth customer receives a free sandwich. If 200 customers show up for the event, how many of the customers will receive both a free drink and a free sandwich?

Unlock the Problem (Real World)

13. Katie is making hair clips to sell at the craft fair. To make each hair clip, she uses 1 barrette and 1 precut ribbon. The barrettes are sold in packs of 12, and the precut ribbons are sold in packs of 9. How many packs of each item does she need to buy to make the least number of hair clips with no supplies left over?

a. What information are you given? _____

_____

b. What problem are you being asked to solve? _____

_____

_____

c. Show the steps you use to solve the problem.

d. Complete the sentences.

The least common multiple of

12 and 9 is _____.

Katie can make _____ hair clips with no supplies left over.

To get 36 barrettes and 36 ribbons, she

needs to buy _____ packs of barrettes

and _____ packs of precut ribbons.

14. **THINK SMARTER** Reptile stickers come in sheets of 6 and fish stickers come in sheets of 9. Antonio buys the same number of both types of stickers and he buys at least 100 of each type. What is the least number of sheets of each type he might buy?

_____

15. **THINK SMARTER** For numbers 15a–15d, choose Yes or No to indicate whether the LCM of the two numbers is 16.

15a. 2, 8   ○ Yes   ○ No

15b. 2, 16   ○ Yes   ○ No

15c. 4, 8   ○ Yes   ○ No

15d. 8, 16   ○ Yes   ○ No

Least Common Multiple

Common Core

COMMON CORE STANDARD—6.NS.B.4
Compute fluently with multi-digit numbers and find common factors and multiples.

Find the LCM.

1. 2, 7

2. 4, 12

3. 6, 9

Multiples of 2: 2, 4, 6, 8, 10, 12, 14
Multiples of 7: 7, 14

LCM: ____14____

LCM: _____

LCM: _____

4. 5, 4

5. 5, 8, 4

6. 12, 8, 24

LCM: _____

LCM: _____

LCM: _____

Write the unknown number for the ▪.

7. 3,   LCM: 21

8. , 7   LCM: 63

9. 10, 5   LCM:

▪ = _____

▪ = _____

▪ = _____

Problem Solving Real World

10. Juanita is making necklaces to give as presents. She plans to put 15 beads on each necklace. Beads are sold in packages of 20. What is the least number of packages she can buy to make necklaces and have no beads left over?

_____

11. Pencils are sold in packages of 10, and erasers are sold in packages of 6. What is the least number of pencils and erasers you can buy so that there is one pencil for each eraser with none left over?

_____

12. **WRITE** ▸*Math* Explain when you would use each method (finding multiples or prime factorization) for finding the LCM and why.

_____

_____

_____

Lesson Check

1. Martha is buying hot dogs and buns for the class barbecue. The hot dogs come in packages of 10. The buns come in packages of 12. What is the least number she can buy of each so that she has exactly the same number of hot dogs and buns? How many packages of each should she buy?

2. Kevin makes snack bags that each contain a box of raisins and a granola bar. Each package of raisins contains 9 boxes. The granola bars come 12 to a package. What is the least number he can buy of each so that he has exactly the same number of granola bars and boxes of raisins? How many packages of each should he buy?

_____

_____

_____

_____

_____

_____

Spiral Review

3. John has 2,456 pennies in his coin collection. He has the same number of pennies in each of 3 boxes. Estimate to the nearest hundred the number of pennies in each box.

4. What is the distance around a triangle that has sides measuring $2\frac{1}{8}$ feet, $3\frac{1}{2}$ feet, and $2\frac{1}{2}$ feet?

_____

_____

5. The 6th grade class collects $1,575. The class wants to give the same amount of money to each of 35 charities. How much will each charity receive?

6. Jean needs $\frac{1}{3}$ cup of walnuts for each serving of salad she makes. She has 2 cups of walnuts. How many servings can she make?

_____

_____

**FOR MORE PRACTICE
GO TO THE
Personal Math Trainer**

Greatest Common Factor

Essential Question How can you find the greatest common factor of two whole numbers?

Common Core **The Number System— 6.NS.B.4**

MATHEMATICAL PRACTICES
MP1, MP4, MP7

A **common factor** is a number that is a factor of two or more numbers. The numbers 16 and 20 have 1, 2, and 4 as common factors.

Factors of 16: 1, 2, 4, 8, 16

Factors of 20: 1, 2, 4, 5, 10, 20

The **greatest common factor**, or **GCF**, is the greatest factor that two or more numbers have in common. The greatest common factor of 16 and 20 is 4.

> **Remember**
> A number that is multiplied by another number to find a product is a factor.
> Factors of 6: 1, 2, 3, 6
> Factors of 9: 1, 3, 9
> Every number has 1 as a factor.

Unlock the Problem Real World

Jim is cutting two strips of wood to make picture frames. The wood strips measure 12 inches and 18 inches. He wants to cut the strips into equal lengths that are as long as possible. Into what lengths should he cut the wood?

12 inches

18 inches

Find the greatest common factor, or GCF, of 12 and 18.

One Way Use a list.

Factors of 12: 1, 2, _____, _____, _____, 12

Factors of 18: 1, _____, _____, _____, _____, _____

The greatest common factor, or GCF, is _____.

Math Talk

MATHEMATICAL PRACTICES ①

Analyze Into what other lengths could Jim cut the wood to obtain equal lengths?

Another Way Use prime factorization.

Write the prime factorization of each number.

$12 = 2 \times$ _____ $\times 3$

$18 =$ _____ $\times 3 \times$ _____

Place the prime factors of the numbers in the appropriate parts of the Venn diagram.

To find the GCF, find the product of the common prime factors.

$2 \times 3 =$ _____   The GCF is _____.

So, Jim should cut the wood into _____-inch lengths.

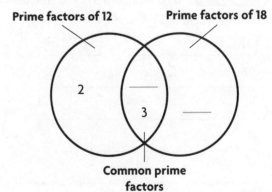

Prime factors of 12   **Prime factors of 18**

2

3

Common prime factors

Distributive Property

Multiplying a sum by a number is the same as multiplying each addend by the number and then adding the products.

$$5 \times (8 + 6) = (5 \times 8) + (5 \times 6)$$

You can use the Distributive Property to express the sum of two whole numbers as a product if the numbers have a common factor.

🔑 Example Use the GCF and the Distributive Property to express 36 + 27 as a product.

Find the GCF of 36 and 27.　　　　GCF: _____

Write each number as the product of the GCF and another factor.

$$36 + 27$$

$$(9 \times \text{_____}) + (9 \times \text{_____})$$

Use the Distributive Property to write 36 + 27 as a product.

$$9 \times (4 + \text{_____})$$

Check your answer.

$$36 + 27 = \text{_____}$$

$$9 \times (4 + \text{_____}) = 9 \times \text{_____} = \text{_____}$$

So, $36 + 27 = \text{_____} \times (\text{_____} + \text{_____})$.

1. Explain two ways to find the GCF of 36 and 27.

_____

_____

2. **MATHEMATICAL PRACTICE ④** Use Diagrams Describe how the figure at the right shows that $36 + 27 = 9 \times (4 + 3)$.

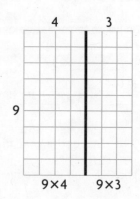

Name _____

1. List the factors of 12 and 20. Circle the GCF.

Factors of 12: _____

Factors of 20: _____

Find the GCF.

2. 16, 18	**3.** 25, 40	✓ **4.** 24, 40	**5.** 14, 35

Use the GCF and the Distributive Property to express the sum as a product.

6. 21 + 28	✓ **7.** 15 + 27	**8.** 40 + 15	**9.** 32 + 20

Math Talk

MATHEMATICAL PRACTICES ①

Analyze Describe how to use the prime factorization of two numbers to find their GCF.

On Your Own

Find the GCF.

10. 8, 25	**11.** 31, 32	**12.** 56, 64	**13.** 150, 275

Use the GCF and the Distributive Property to express the sum as a product.

14. 24 + 30	**15.** 49 + 14	**16.** 63 + 81	**17.** 60 + 12

18. **MATHEMATICAL PRACTICE ①** **Describe** the difference between the LCM and the GCF of two numbers.

_____

Problem Solving • Applications ⟨Real World⟩

Use the table for 19–22. Teachers at the Scott School of Music teach only one instrument in each class. No students take classes for more than one instrument.

19. Francisco teaches group lessons to all of the violin and viola students at the Scott School of Music. All of his classes have the same number of students. What is the greatest number of students he can have in each class?

_____

Scott School of Music	
Instrument	**Number of Students**
Bass	20
Cello	27
Viola	30
Violin	36

20. GO DEEPER Amanda teaches all of the bass and viola students. All her classes have the same number of students. Each class has the greatest possible number of students. How many of these classes does she teach?

_____

21. THINK SMARTER Mia teaches jazz classes. She has 9 students in each class, and she teaches all the classes for two of the instruments. Which two instruments does she teach, and how many students are in her classes?

_____

22. ◼ WRITE ▸Math Explain how you could use the GCF and the Distributive Property to express the sum of the number of bass students and the number of violin students as a product.

_____

_____

_____

23. THINK SMARTER The prime factorization of each number is shown.

$6 = 2 \times 3$
$12 = 2 \times 2 \times 3$

Using the prime factorization, complete the Venn diagram and write the GCF of 6 and 12.

GCF = _____

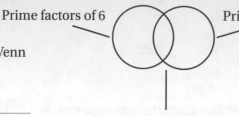

Prime factors of 6   Prime factors of 12

Common prime factors

Greatest Common Factor

Common
Core

COMMON CORE STANDARD—6.NS.B.4
Compute fluently with multi-digit numbers and find common factors and multiples.

List the common factors. Circle the greatest common factor.

1. 25 and 10

2. 36 and 90

3. 45 and 60

1, ⑤ _____

_____

_____

Find the GCF.

4. 14, 18

5. 6, 48

6. 16, 100

_____

_____

_____

Use the GCF and the Distributive Property to express the sum as a product.

7. 20 + 35

8. 18 + 27

9. 64 + 40

_____

_____

_____

Problem Solving · Real World

10. Jerome is making prizes for a game at the school fair. He has two bags of different pins, one with 15 square pins and one with 20 round pins. Every prize will have one kind of pin. Each prize will have the same number of pins. What is the greatest number of pins Jerome can put in each prize?

11. There are 24 sixth graders and 40 seventh graders. Mr. Chan wants to divide both grades into groups of equal size, with the greatest possible number of students in each group. How many students should be in each group?

_____

_____

12. **WRITE** ▸*Math* Write a short paragraph to explain how to use prime factorization and the Distributive Property to express the sum of two whole numbers as a product.

_____

_____

Lesson Check (6.NS.B.4)

1. There are 15 boys and 10 girls in Miss Li's class. She wants to group all the students so that each group has the same number of boys and the same number of girls. What is the greatest number of groups she can have?

_____

2. A pet shop manager wants the same number of birds in each cage. He wants to use as few cages as possible, but can only have one type of bird in each cage. If he has 42 parakeets and 18 canaries, how many birds will he put in each cage?

_____

Spiral Review (5.NBT.A.1, 5.NBT.B.6, 5.NF.B.7c, 6.NS.B.2)

3. There are 147 people attending a dinner party. If each table can seat 7 people, how many tables are needed for the dinner party?

_____

4. Sammy has 3 pancakes. He cuts each one in half. How many pancake halves are there?

_____

5. The Cramer Company had a profit of $8,046,890 and the Coyle Company had a profit of $8,700,340 last year. Which company had the greater profit?

_____

6. There are 111 guests attending a party. There are 15 servers. Each server has the same number of guests to serve. Jess will serve any extra guests. How many guests will Jess be serving?

_____

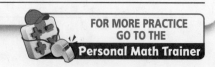

FOR MORE PRACTICE
GO TO THE
Personal Math Trainer

Problem Solving • Apply the Greatest Common Factor

Essential Question How can you use the strategy *draw a diagram* to help you solve problems involving the GCF and the Distributive Property?

 Common Core **The Number System—6.NS.B.4**
MATHEMATICAL PRACTICES
MP1, MP5, MP6

🔑 Unlock the Problem

A trophy case at Riverside Middle School holds 18 baseball trophies and 24 soccer trophies. All shelves hold the same number of trophies. Only one sport is represented on each shelf. What is the greatest number of trophies that can be on each shelf? How many shelves are there for each sport?

Use the graphic organizer to help you solve the problem.

Read the Problem	Solve the Problem
What do I need to find? I need to find _____ _____ _____	Total trophies = baseball + soccer 18 + 24 Find the GCF of 18 and 24. GCF: _____ Write each number as the product of the GCF and another factor. 18 + 24 (6 × _____) + (6 × _____) Use the Distributive Property to write 18 + 24 as a product. 6 × (_____ + _____)
What information do I need to use? I need to use _____ _____	
How will I use the information? I can find the GCF of _____ and use it to draw a diagram representing the _____ of the trophy case.	Use the product to draw a diagram of the trophy case. Use B's to represent baseball trophies. Use S's to represent soccer trophies. B B B B B B S S S S S S

 Math Talk MATHEMATICAL PRACTICES ⑤

Use Tools Explain how the Distributive Property helped you solve the problem.

So, there are _____ trophies on each shelf. There are _____ shelves of

baseball trophies and _____ shelves of soccer trophies.

🔒 Try Another Problem

Delia is bagging 24 onion bagels and 16 plain bagels for her bakery
customers. Each bag will hold only one type of bagel. Each bag will hold the
same number of bagels. What is the greatest number of bagels she can put in
each bag? How many bags of each type of bagel will there be?

Use the graphic organizer to help you solve the problem.

Read the Problem	Solve the Problem
What do I need to find?	
What information do I need to use?	
How will I use the information?	

So, there will be _____ bagels in each bag. There will be

_____ bags of onion bagels and _____ bags of plain bagels.

- **MATHEMATICAL PRACTICE ⑥** **Explain** how knowing that the GCF of 24 and 16 is 8 helped
you solve the bagel problem.

_____

_____

Name _____

Unlock the Problem

✓ Circle important facts.
✓ Check to make sure you answered the question.
✓ Check your answer.

1. Toby is packaging 21 baseball cards and 12 football cards to sell at a swap meet. Each packet will have the same number of cards. Each packet will have cards for only one sport. What is the greatest number of cards he can place in each packet? How many packets will there be for each sport?

   First, find the GCF of 21 and 12.

   _____

   Next, use the Distributive Property to write 21 + 12 as a product, with the GCF as one of the factors.

   _____

   So, there will be _____ packets of baseball cards and

   _____ packets of football cards. Each packet will

   contain _____ cards.

WRITE *Math*
Show Your Work

2. **THINK SMARTER** What if Toby had decided to keep one baseball card for himself and sell the rest? How would your answers to the previous problem have changed?

   _____

   _____

   _____

3. Melissa bought 42 pine seedlings and 30 juniper seedlings to plant in rows on her tree farm. She wants each row to have the same number of seedlings. She wants only one type of seedling in each row. What is the greatest number of seedlings she can plant in each row? How many rows of each type of tree will there be?

   _____

   _____

On Your Own

4. **MATHEMATICAL PRACTICE ①** **Make Sense of Problems** A drum and bugle marching band has 45 members who play bugles and 27 members who play drums. When they march, each row has the same number of players. Each row has only bugle players or only drummers. What is the greatest number of players there can be in each row? How many rows of each type of player can there be?

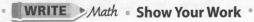

_____

_____

5. **THINK SMARTER** The "color guard" of a drum and bugle band consists of members who march with flags, hoops, and other props. How would your answers to Exercise 4 change if there were 21 color guard members marching along with the bugle players and drummers?

_____

_____

_____

6. **GO DEEPER** If you continue the pattern below so that you write all of the numbers in the pattern less than 500, how many even numbers will you write?

$$4, 9, 14, 19, 24, 29...$$

_____

Personal Math Trainer

7. **THINK SMARTER +** Mr. Yaw's bookcase holds 20 nonfiction books and 15 fiction books. Each shelf holds the same number of books and contains only one type of book. How many books will be on each shelf if each shelf has the **greatest** possible number of books? Show your work.

_____

_____

_____

Problem Solving • Apply the Greatest Common Factor

 COMMON CORE STANDARD—6.NS.B.4
Compute fluently with multi-digit numbers and find common factors and multiples.

Read the problem and solve.

1. Ashley is bagging 32 pumpkin muffins and 28 banana muffins for some friends. Each bag will hold only one type of muffin. Each bag will hold the same number of muffins. What is the greatest number of muffins she can put in each bag? How many bags of each type of muffin will there be?

   GCF: 4

   $32 = 4 \times 8$

   $28 = 4 \times 7$

   $32 + 28 = 4 \times (8 + 7)$

   So, there will be __8__ bags of pumpkin muffins and __7__ bags of banana muffins,

   with __4__ muffins in each bag.

2. Patricia is separating 16 soccer cards and 22 baseball cards into groups. Each group will have the same number of cards, and each group will have only one kind of sports card. What is the greatest number of cards she can put in each group? How many groups of each type will there be?

   _____

3. Bryan is setting chairs in rows for a graduation ceremony. He has 50 black chairs and 60 white chairs. Each row will have the same number of chairs, and each row will have the same color chair. What is the greatest number of chairs that he can fit in each row? How many rows of each color chair will there be?

   _____

4. A store clerk is bagging spices. He has 18 teaspoons of cinnamon and 30 teaspoons of nutmeg. Each bag needs to contain the same number of teaspoons, and each bag can contain only one spice. What is the maximum number of teaspoons of spice the clerk can put in each bag? How many bags of each spice will there be?

   _____

5. **WRITE** ▸*Math* Write a problem in which you need to put as many of two different types of objects as possible into equal groups. Then use the GCF, Distributive Property, and a diagram to solve your problem.

_____

Lesson Check (6.NS.B.4)

1. Fred has 36 strawberries and 42 blueberries. He wants to use them to garnish desserts so that each dessert has the same number of berries, but only one type of berry. He wants as much fruit as possible on each dessert. How many berries will he put on each dessert? How many desserts with each type of fruit will he have?

_____

_____

_____

2. Dolores is arranging coffee mugs on shelves in her shop. She wants each shelf to have the same number of mugs. She only wants one color of mug on each shelf. If she has 49 blue mugs and 56 red mugs, what is the greatest number she can put on each shelf? How many shelves does she need for each color?

_____

_____

_____

Spiral Review (5.NF.A.1, 5.NF.A.2, 6.NS.B.4)

3. A rectangle is $3\frac{1}{3}$ feet long and $2\frac{1}{3}$ feet wide. What is the distance around the rectangle?

_____

4. Lowell bought $4\frac{1}{4}$ pounds of apples and $3\frac{3}{5}$ pounds of oranges. How many pounds of fruit did Lowell buy?

_____

5. How much heavier is a $9\frac{1}{8}$ pound box than a $2\frac{5}{6}$ pound box?

_____

6. The combination of Clay's locker is the prime factors of 102 in order from least to greatest. What is the combination of Clay's locker?

_____

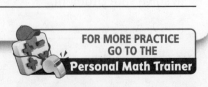
FOR MORE PRACTICE
GO TO THE
Personal Math Trainer

 **Mid-Chapter Checkpoint**

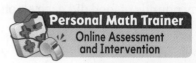
Personal Math Trainer
Online Assessment
and Intervention

Vocabulary

Choose the best term from the box to complete the sentence.

1. The _____ of two numbers is greater than or equal to the numbers. (p. 17)

2. The _____ of two numbers is less than or equal to the numbers. (p. 23)

Concepts and Skills

Estimate. Then find the quotient. Write the remainder, if any, with an r. (6.NS.B.2)

3. $2,800 \div 25$

4. $19,129 \div 37$

5. $32,111 \div 181$

_____

_____

_____

Find the prime factorization. (6.NS.B.4)

6. 44

7. 36

8. 90

_____

_____

_____

Find the LCM. (6.NS.B.4)

9. 8, 10

10. 4, 14

11. 6, 9

_____

_____

_____

Find the GCF. (6.NS.B.4)

12. 16, 20

13. 8, 52

14. 36, 54

_____

_____

_____

15. A zookeeper divided 2,440 pounds of food equally among 8 elephants. How many pounds of food did each elephant receive? (6.NS.B.2)

---

16. DVD cases are sold in packages of 20. Padded mailing envelopes are sold in packets of 12. What is the least number of cases and envelopes you could buy so that there is one case for each envelope with none left over? (6.NS.B.4)

---

17. GO DEEPER Max bought two deli sandwich rolls measuring 18 inches and 30 inches. He wants them to be cut into equal sections that are as long as possible. Into what lengths should the rolls be cut? How many sections will there be in all? (6.NS.B.4)

---

18. Susan is buying supplies for a party. If spoons only come in bags of 8 and forks only come in bags of 6, what is the least number of spoons and the least number of forks she can buy so that she has the same number of each? (6.NS.B.4)

---

19. GO DEEPER Tina is placing 30 roses and 42 tulips in vases for table decorations in her restaurant. Each vase will hold the same number of flowers. Each vase will have only one type of flower. What is the greatest number of flowers she can place in each vase? If Tina has 24 tables in her restaurant, how many flowers can she place in each vase? (6.NS.B.4)

Add and Subtract Decimals

Essential Question How do you add and subtract multi-digit decimals?

 Common Core **The Number System—6.NS.B.3**

MATHEMATICAL PRACTICES
MP3, MP6, MP7

CONNECT The place value of a digit in a number shows the value of the digit. The number 2.358 shows 2 ones, 3 tenths, 5 hundredths, and 8 thousandths.

Place Value						
Thousands	Hundreds	Tens	Ones	Tenths	Hundredths	Thousandths
			2	3	5	8

Unlock the Problem (Real World)

Amanda and three of her friends volunteer at the local animal shelter. One of their jobs is to weigh the puppies and kittens and chart their growth. Amanda's favorite puppy weighed 2.358 lb last month. If it gained 1.08 lb, how much does it weigh this month?

- How do you know whether to add or subtract the weights given in the problem?

_____

_____

 Add 2.358 + 1.08.

Estimate the sum. _____ + _____ = _____

$$\begin{array}{r} 2.358 \\ + 1.08 \\ \hline \end{array}$$

Compare your estimate with the sum. Since the estimate,

_____, is close to _____, the answer is reasonable.

So, the puppy weighs _____ lb this month.

1. **MATHEMATICAL PRACTICE 7** **Look for Structure** Is it necessary to write a zero after 1.08 to find the sum? Explain.

_____

2. Explain how place value can help you add decimals.

_____

🔓 Example 1

A bee hummingbird, the world's smallest bird, has a mass of 1.836 grams. A new United States nickel has a mass of 5 grams. What is the difference in grams between the mass of a nickel and the mass of a bee hummingbird?

Subtract 5 − 1.836.

Estimate the difference. _____ − _____ = _____

Think: 5 = 5._____

Subtract the thousandths first.

Then subtract the hundredths, tenths, and ones.

Regroup as needed.

$$\begin{array}{r} 5. \\ -1.836 \\ \hline \end{array}$$

Bee hummingbird

U.S. Nickel

Compare your estimate with the difference. Since the estimate,

_____, is close to _____, the answer is reasonable.

So, the mass of a new nickel is _____ grams more than the mass of a bee hummingbird.

Math Talk

MATHEMATICAL PRACTICES ⑥

Explain how to use inverse operations to check your answer to 5 − 1.836.

🔓 Example 2 Evaluate (6.5 − 1.97) + 3.461 using the order of operations.

Write the expression.

$$(6.5 - 1.97) + 3.461$$

Perform operations in parentheses.

$$\begin{array}{r} 6.50 \\ -1.97 \\ \hline \end{array}$$

Add.

$$\begin{array}{r} \\ + 3.461 \\ \hline \end{array}$$

Math Talk

MATHEMATICAL PRACTICES ⑦

Look for Structure Describe how adding and subtracting decimals is like adding and subtracting whole numbers.

So, the value of the expression is _____.

Name _____

1. Find $3.42 - 1.9$.

Estimate.

_____ − _____ = _____

Subtract the _____ first.

$$\begin{array}{r} 3.42 \\ -1.90 \\ \hline \end{array}$$

Estimate. Then find the sum or difference.

✓ **2.** $2.3 + 5.68 + 21.047$

✓ **3.** $33.25 - 21.463$

4. Evaluate
$(8.54 + 3.46) - 6.749$.

_____ _____ _____

Math Talk   MATHEMATICAL PRACTICES ⑥

Explain why it is important to align the decimal points when you add or subtract decimals.

On Your Own

Estimate. Then find the sum or difference.

5. $57.08 + 34.71$

6. $20.11 - 13.27$

7. $62 - 9.817$

8. $35.1 + 4.89$

_____ _____ _____ _____

Practice: Copy and Solve Evaluate using the order of operations.

9. $8.01 - (2.2 + 4.67)$

10. $54 + (9.2 - 1.413)$

11. $21.3 - (19.1 - 3.22)$

12. MATHEMATICAL PRACTICE ③ **Make Arguments** A student evaluated $19.1 + (4.32 + 6.9)$ and got 69.2. How can you use estimation to convince the student that this answer is not reasonable?

_____

13. *THINK SMARTER* Lynn paid $4.75 for cereal, $8.96 for chicken, and $3.25 for soup. Show how she can use properties and compatible numbers to evaluate $(4.75 + 8.96) + 3.25$ to find the total cost.

_____

14. **THINK SMARTER** For numbers 14a–14d, select True or False for each equation.

14a. $3.76 + 2.7 = 6.46$ ○ True   ○ False

14b. $4.14 + 1.8 = 4.32$ ○ True   ○ False

14c. $2.01 - 1.33 = 0.68$ ○ True   ○ False

14d. $51 - 49.2 = 1.8$ ○ True   ○ False

Connect to Science

Comparing Eggs

Different types of birds lay eggs of different sizes. Small birds lay eggs that are smaller than those that are laid by larger birds. The table shows the average lengths and widths of five different birds' eggs.

Average Dimensions of Bird Eggs		
Bird	Length (m)	Width (m)
Canada Goose	0.086	0.058
Hummingbird	0.013	0.013
Raven	0.049	0.033
Robin	0.019	0.015
Turtledove	0.031	0.023

Canada Goose

Use the table for 15–17.

15. What is the difference in average length between the longest egg and the shortest egg?

_____

16. **GO DEEPER** Which egg has a width that is eight thousandths of a meter shorter than its length?

_____

17. **THINK SMARTER** How many robin eggs, laid end to end, would be about equal in length to two raven eggs? Justify your answer.

_____

_____

Add and Subtract Decimals

Common Core   COMMON CORE STANDARD—6.NS.B.3
Compute fluently with multi-digit numbers and
find common factors and multiples.

Estimate. Then find the sum or difference.

1. $43.53 + 27.67$

$40 + 30 = 70$

$$\begin{array}{r} 43.53 \\ + \ 27.67 \\ \hline 71.20 \end{array}$$

2. $17 + 3.6 + 4.049$

3. $3.49 - 2.75$

4. $5.07 - 2.148$

5. $3.92 + 16 + 0.085$

6. $41.98 + 13.5 + 27.338$

Evaluate using the order of operations.

7. $8.4 + (13.1 - 0.6)$

8. $34.7 - (12.07 + 4.9)$

9. $(32.45 - 4.8) - 2.06$

Problem Solving (Real World)

10. The average annual rainfall in Clearview is 38 inches. This year, 29.777 inches fell. How much less rain fell this year than falls in an average year?

11. At the theater, the Worth family spent $18.00 on adult tickets, $16.50 on children's tickets, and $11.75 on refreshments. How much did they spend in all?

12. **WRITE** ▸ *Math* Write a word problem that involves adding or subtracting decimals. Include the solution.

_____

_____

_____

Lesson Check (6.NS.B.3)

1. Alden fills his backpack with 0.45 kg of apples, 0.18 kg of cheese, and a water bottle that weighs 1.4 kg. How heavy are the contents of his backpack?

   _____

   _____

2. Gabby plans to hike 6.3 kilometers to see a waterfall. She stops to rest after hiking 4.75 kilometers. How far does she have left to hike?

   _____

   _____

Spiral Review (5.NBT.B.5, 5.NBT.B.6, 6.NS.B.4)

3. A 6-car monorail train can carry 78 people. If one train makes 99 trips during the day, what is the greatest number of people the train can carry in one day?

   _____

4. An airport parking lot has 2,800 spaces. If each row has 25 spaces, how many rows are there?

   _____

5. Evan brought 6 batteries that cost $10 each and 6 batteries that cost $4 each. The total cost was the same as he would have spent buying 6 batteries that cost $14 each. So, $6 \times \$14 = (6 \times 10) + (6 \times 4)$. What property does the equation illustrate?

   _____

6. Cups come in packages of 12 and lids come in packages of 15. What is the least number of cups and lids that Corrine can buy if she wants to have the same number of cups and lids?

   _____

FOR MORE PRACTICE
GO TO THE
Personal Math Trainer

Multiply Decimals

Essential Question How do you multiply multi-digit decimals?

Common Core The Number System—
6.NS.B.3

MATHEMATICAL PRACTICES
MP2, MP6, MP8

Unlock the Problem Real World

Last summer Rachel worked 38.5 hours per week at a grocery store. She earned $9.70 per hour. How much did she earn in a week?

- **How can you estimate the product?**

_____

_____

Multiply $9.70 × 38.5.

First estimate the product. $10 × 40 = _____

You can use the estimate to place the decimal in a product.

$$\begin{array}{r} \$9.70 \\ \times\,38.5 \\ \hline \end{array}$$

Multiply as you would with whole numbers.

The estimate is about $ _____,

so the decimal point should be

placed after $_____.

Since the estimate, _____, is close to _____, the answer is reasonable.

So, Rachel earned _____ per week.

---

1. Explain how your estimate helped you know where to place the decimal in the product.

_____

_____

Try This! **What if** Rachel gets a raise of $1.50 per hour? How much will she earn when she works 38.5 hours?

Counting Decimal Places Another way to place the decimal in a
product is to add the numbers of decimal places in the factors.

🔒 **Example 1** Multiply 0.084 × 0.096.

$$0.084$$
$$\times\, 0.096$$

_____ decimal places

_____ decimal places

Multiply as you would with whole numbers.

+ _____

_____ + _____, or _____ decimal places

🔒 **Example 2** Evaluate 0.35 × (0.48 + 1.24) using the order of operations.

Write the expression.

$$0.35 \times (0.48 + 1.24)$$

Perform operations in parentheses.

$$0.35 \times \underline{\hspace{2cm}}$$

Multiply.

$$0.35$$ _____ decimal places

$$\times$$ _____ decimal places

+

_____ + _____, or _____ decimal places

So, the value of the expression is _____.

MATHEMATICAL PRACTICES ②

Reasoning Is the product of 0.5 and 3.052 greater than or less than 3.052?

---

2. **MATHEMATICAL PRACTICE ⑧ Use Repeated Reasoning** Look for a pattern. Explain.

0.645 × 1 = 0.645

0.645 × 10 = 6.45   The decimal point moves _____ place to the right.

0.645 × 100 = _____   The decimal point moves _____ places to the right.

0.645 × 1,000 = _____   The decimal point moves _____ places to the right.

_____

_____

44

Name _____

Estimate. Then find the product.

1. 12.42×28.6

_____ $\times$ _____ = _____

$$\begin{array}{r} 12.42 \\ \times\ 28.6 \\ \hline \end{array}$$

Estimate.

Think: The estimate is about _____, so the decimal point should be placed after _____.

✓**2.** 32.5×7.4

_____

MATHEMATICAL PRACTICE ⑥ **Attend to Precision** **Algebra** Evaluate using the order of operations.

3. $0.24 \times (7.3 + 2.1)$

✓**4.** $0.075 \times (9.2 - 0.8)$

5. $2.83 + (0.3 \times 2.16)$

On Your Own

Estimate. Then find the product.

6. 29.14×5.2

7. 6.95×12

8. 0.055×1.82

MATHEMATICAL PRACTICE ⑥ **Attend to Precision** **Algebra** Evaluate using the order of operations.

9. $(3.62 \times 2.1) - 0.749$

10. $5.8 - (0.25 \times 1.5)$

11. $(0.83 + 1.27) \times 6.4$

---

12. GO DEEPER Jamal is buying ingredients to make a large batch of granola to sell at a school fair. He buys 3.2 pounds of walnuts for $4.40 per pound and 2.4 pounds of cashews for $6.25 per pound. How much change will he receive if he pays with two $20 bills?

_____

Unlock the Problem

The table shows some currency exchange rates for 2009.

Major Currency Exchange Rates in 2009				
Currency	U.S. Dollar	Japanese Yen	European Euro	Canadian Dollar
U.S. Dollar	1	88.353	0.676	1.052
Japanese Yen	0.011	1	0.008	0.012
European Euro	1.479	130.692	1	1.556
Canadian Dollar	0.951	83.995	0.643	1

Denominations of Euro

13. **THINK SMARTER** When Cameron went to Canada in 2007, he exchanged 40 U.S. dollars for 46.52 Canadian dollars. If Cameron exchanged 40 U.S. dollars in 2009, did he receive more or less than he received in 2007? How much more or less?

a. What do you need to find?

_____

_____

b. How will you use the table to solve the problem?

_____

_____

_____

c. Complete the sentences.

40 U.S. dollars were worth _____ Canadian dollars in 2009.

So, Cameron would receive _____

_____ Canadian dollars in 2009.

Personal Math Trainer

14. **THINK SMARTER +** At a convenience store, the Jensen family puts 12.4 gallons of gasoline in their van at a cost of $3.80 per gallon. They also buy 4 water bottles for $1.99 each, and 2 snacks for $1.55 each. Complete the table to find the cost for each item.

Item	Calculation	Cost
Gasoline	12.4 × $3.80	
Water bottles	4 × $1.99	
Snacks	2 × $1.55	

Mrs. Jensen says the total cost for everything before tax is $56.66. Do you agree with her? Explain why or why not.

_____

Common Core

COMMON CORE STANDARD—6.NS.B.3
Compute fluently with multi-digit numbers and find common factors and multiples.

Estimate. Then find the product.

1. 5.69×7.8

$6 \times 8 = 48$

$$\begin{array}{r} 5.69 \\ \times\ 7.8 \\ \hline 4552 \\ 39830 \\ \hline 44.382 \end{array}$$

2. 3.92×0.051

3. 2.365×12.4

4. 305.08×1.5

Evaluate the expression using the order of operations.

5. $(61.8 \times 1.7) + 9.5$

6. $205 - (35.80 \times 5.6)$

7. $1.9 \times (10.6 - 2.17)$

Problem Solving Real World

8. Blaine exchanges $100 for yen before going to Japan. If each U.S. dollar is worth 88.353 yen, how many yen should Blaine receive?

9. A camera costs 115 Canadian dollars. If each Canadian dollar is worth 0.952 U.S. dollars, how much will the camera cost in U.S. dollars?

10. **WRITE** *Math* Explain how to mentally multiply a decimal number by 100.

Lesson Check (6.NS.B.3)

Estimate each product. Then find the exact product for each question.

1. A gallon of water at room temperature weighs about 8.35 pounds. Lena puts 4.5 gallons in a bucket. How much does the water weigh?

2. Shawn's rectangular mobile home is 7.2 meters wide and 19.5 meters long. What is its area?

Spiral Review (5.OA.A.1, 6.NS.B.2, 6.NS.B.4)

3. Last week, a store sold laptops worth a total of $3,885. Each laptop cost $555. How many laptops did the store sell last week?

4. Kyle drives his truck 429 miles on 33 gallons of gas. How many miles can Kyle drive on 1 gallon of gas?

5. Seven busloads each carrying 35 students arrived at the game, joining 23 students who were already there. Evaluate the expression $23 + (7 \times 35)$ to find the total number of students at the game.

6. A store is giving away a $10 coupon to every 7th person to enter the store and a $25 coupon to every 18th person to enter the store. Which person will be the first to get both coupons?

© Houghton Mifflin Harcourt Publishing Company

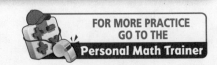

FOR MORE PRACTICE
GO TO THE
Personal Math Trainer

Divide Decimals by Whole Numbers

Essential Question How do you divide decimals by whole numbers?

 **The Number System—
6.NS.B.3**

**MATHEMATICAL PRACTICES
MP1, MP5, MP6, MP8**

🔑 Unlock the Problem

Dan opened a savings account at a bank to save for a new snowboard. He earned $3.48 interest on his savings account over a 3-month period. What was the average amount of interest Dan earned per month on his savings account?

🔒 **Divide $3.48 ÷ 3.**

First estimate. 3 ÷ 3 = _____

```
     1.
3)3.48
 -3 ↓
   04
 - 3 ↓
   18
  -18
    0
```

Think: 3.48 is shared among 3 groups.

Divide the ones. Place a decimal point after the ones place in the quotient.

Divide the tenths and then the hundredths. When the remainder is zero and there are no more digits in the dividend, the division is complete.

> **Remember**
>
> Quotient
> ↓
> 1.23
> Divisor → 2)2.46 ← Dividend

Check your answer.

```
  $
×    3
$3.48
```

Multiply the quotient by the divisor to check your answer.

So, Dan earned an average of _____ in interest per month.

Math Talk

MATHEMATICAL PRACTICES ⑥

Explain how you know your answer is reasonable.

1. **MATHEMATICAL PRACTICE ①** **Analyze Relationships** What if the same amount of interest was gained over 4 months? Explain how you would solve the problem.

_____

Example Divide 42.133 ÷ 7.

First estimate. 42 ÷ 7 = _____

```
      6.0▨
  7)42.133
   -42
     01
     -0
      13
      -7
       63
       -▨
        ▨
```

Think: 42.133 is shared among 7 groups.

Divide the ones. Place a decimal point after the ones place in the quotient.

Divide the tenths. Since 1 tenth cannot be shared among 7 groups, write a zero in the quotient. Regroup the 1 tenth as 10 hundredths. Now you have 13 hundredths.

Continue to divide until the remainder is zero and there are no more digits in the dividend.

Check your answer.

```
   6.019
 ×     7
   ▨▨▨▨
```

Multiply the quotient by the divisor to check your answer.

So, 42.133 ÷ 7 = _____.

2. Explain how you know which numbers to multiply when checking your answer.

_____

_____

 **Share and Show** MATH BOARD

1. Estimate 24.186 ÷ 6. Then find the quotient. Check your answer.

Estimate. _____ ÷ _____ = _____

Think: Place a decimal point after the ones place in the quotient.

```
 6)24.186          ×      6
```

50

Name _____

Estimate. Then find the quotient.

2. $7\overline{)\$17.15}$ ✓ **3.** $4\overline{)1.068}$ **4.** $12\overline{)60.84}$ ✓ **5.** $18.042 \div 6$

Math Talk   MATHEMATICAL PRACTICES ⑧

Generalize Explain how you know where to place the decimal point in the quotient when dividing a decimal by a whole number.

On Your Own

Estimate. Then find the quotient.

6. $\$21.24 \div 6$

7. $28.63 \div 7$

8. $1.505 \div 35$

9. $0.108 \div 18$

MATHEMATICAL PRACTICE ⑥ **Attend to Precision Algebra** Evaluate using the order of operations.

10. $(3.11 + 4.0) \div 9$

11. $(6.18 - 1.32) \div 3$

12. $(18 - 5.76) \div 6$

13. MATHEMATICAL PRACTICE ⑤ **Use Appropriate Tools** Find the length of a dollar bill to the nearest tenth of a centimeter. Then show how to use division to find the length of the bill when it is folded in half along the portrait of George Washington.

_____

14. GO DEEPER Emilio bought 5.65 pounds of green grapes and 3.07 pounds of red grapes. He divided the grapes equally into 16 bags. If each bag of grapes has the same weight, how much does each bag weigh?

_____

Problem Solving • Applications Real World

Pose a Problem

15. **THINK SMARTER** This table shows the average height in inches for girls and boys at ages 8, 10, 12, and 14 years.

Average Height (in.)				
	Age 8	**Age 10**	**Age 12**	**Age 14**
Girls	50.75	55.50	60.50	62.50
Boys	51.00	55.25	59.00	65.20

To find the average growth per year for girls from age 8 to age 12, Emma knew she had to find the amount of growth between age 8 and age 12, then divide that number by the number of years between age 8 and age 12.

Emma used this expression: $(60.50 - 50.75) \div 4$

She evaluated the expression using the order of operations.

Write the expression.	$(60.50 - 50.75) \div 4$
Perform operations in parentheses.	$9.75 \div 4$
Divide.	2.4375

So, the average annual growth for girls ages 8 to 12 is 2.4375 inches.

Write a new problem using the information in the table for the average height for boys. Use division in your problem.

Pose a Problem	**Solve Your Problem**
_____ _____ _____	

16. **THINK SMARTER** The table shows the number of books each of three friends bought and the cost. On average, which friend spent the most per book? Use numbers and words to explain your answer.

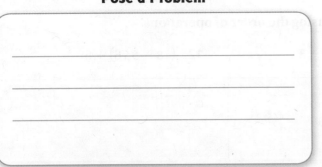

Friend	Number of books Purchased	Total Cost (in dollars)	Average Cost (in dollars)
Joyce	1	$10.95	
Nabil	2	$40.50	
Kenneth	3	$51.15	

_____

_____

Divide Decimals by Whole Numbers

Common Core   **COMMON CORE STANDARD—6.NS.B.3**
Compute fluently with multi-digit numbers and find common factors and multiples.

Estimate. Then find the quotient.

1. $1.284 \div 12$ **2.** $9\overline{)2.43}$ **3.** $25.65 \div 15$ **4.** $12\overline{)2.436}$

$1.2 \div 12 = 0.1$

$$
\begin{array}{r}
0.107 \\
12\overline{)1.284} \\
\underline{-12} \\
-8 \\
\underline{0} \\
84 \\
\underline{-84} \\
0
\end{array}
$$

Evaluate using the order of operations.

5. $(8 - 2.96) \div 3$ **6.** $(7.772 - 2.38) \div 8$ **7.** $(53.2 + 35.7) \div 7$

Problem Solving Real World

8. Jake earned $10.44 interest on his savings account for an 18-month period. What was the average amount of interest Jake earned on his savings account per month?

9. Gloria worked for 6 hours a day for 2 days at the bank and earned $114.24. How much did she earn per hour?

10. **WRITE** ▶ *Math* Explain the importance of correctly placing the decimal point in the quotient of a division problem.

_____

_____

Lesson Check (6.NS.B.3)

Estimate each quotient. Then find the exact quotient for each question.

1. Ron divided 67.6 fluid ounces of orange juice evenly among 16 glasses. How much did he pour into each glass?

_____

_____

_____

_____

2. The cost of a $12.95 pizza was shared evenly by 5 friends. How much did each person pay?

_____

_____

_____

_____

Spiral Review (5.NBT.A.1, 5.NBT.B.6, 6.NS.B.2, 6.NS.B.4)

3. What is the value of the digit 6 in 968,743,220?

_____

4. The Tama, Japan, monorail carries 92,700 riders each day. If the monorail runs 18 hours each day, what is the average number of passengers riding each hour?

_____

5. Ray paid $812 to rent music equipment that costs $28 per hour. How many hours did he have the equipment?

_____

6. Jan has 35 teaspoons of chocolate cocoa mix and 45 teaspoons of french vanilla cocoa mix. She wants to put the same amount of mix into each jar, and she only wants one flavor of mix in each jar. She wants to fill as many jars as possible. How many jars of french vanilla cocoa mix will Jan fill?

_____

**FOR MORE PRACTICE
GO TO THE
Personal Math Trainer**

Name _____

Divide with Decimals

Essential Question How do you divide whole numbers and decimals by decimals?

Common Core
The Number System—
6.NS.B.3
MATHEMATICAL PRACTICES
MP1, MP6, MP8

CONNECT Find each quotient to discover a pattern.

$4 \div 2 =$ _____

$40 \div 20 =$ _____

$400 \div 200 =$ _____

When you multiply both the dividend and the divisor by the same

power of _____, the quotient is the _____. You can use this fact to help you divide decimals.

Unlock the Problem

Tami is training for a triathlon. In a triathlon, athletes compete in three events: swimming, cycling, and running. She cycled 66.5 miles in 3.5 hours. If she cycled at a constant speed, how far did she cycle in 1 hour?

 Divide 66.5 ÷ 3.5.

Estimate using compatible numbers.

$60 \div 3 =$ _____

STEP 1

Make the divisor a whole number by multiplying the divisor and dividend by 10.

$3.5 \overline{)66.5}$

Think: $3.5 \times 10 = 35$ $66.5 \times 10 = 665$

STEP 2

Divide.

So, Tami cycled _____ in 1 hour.

> **Remember**
> Compatible numbers are pairs of numbers that are easy to compute mentally.

MATHEMATICAL PRACTICE ① **Evaluate Reasonableness** Explain whether your answer is reasonable.

_____

🔒 Example 1 Divide 17.25 ÷ 5.75. Check.

STEP 1

Make the divisor a whole number by multiplying the divisor and dividend by _____.

5.75 × _____ = _____

17.25 × _____ = _____

$$5.75\overline{)17.25}$$

STEP 2

Divide.

$$575\overline{)1,725}$$

STEP 3

Check.

So, 17.25 ÷ 5.75 = _____.

×

🔒 Example 2 Divide 37.8 ÷ 0.14.

STEP 1

Make the divisor a whole number by multiplying the divisor and dividend by _____.

_____ × _____ = _____

_____ × _____ = _____

$$0.14\overline{)37.80}$$

Think: Write a zero to the right of the dividend so that you can move the decimal point.

> ⚠️ **ERROR Alert**
>
> Be careful to move the decimal point in the dividend the same number of places that you moved the decimal point in the divisor.

STEP 2

Divide.

$$14\overline{)3,780}$$

So, 37.8 ÷ 0.14 = _____.

Math Talk

MATHEMATICAL PRACTICES ⑥

Explain How can you check that the quotient is reasonable? How can you check that it is accurate?

Name _____

Share and Show MATH BOARD

1. Find the quotient.

$$14.8\overline{)99.456}$$

Think: Make the divisor a whole number by

multiplying the divisor and dividend by _____.

Estimate. Then find the quotient.

2. $10.80 ÷ $1.35

3. 26.4 ÷ 1.76

4. $8.7\overline{)53.07}$

Math Talk

MATHEMATICAL PRACTICES ⑧

Generalize Explain how you know how many places to move the decimal point in the divisor and the dividend.

On Your Own

Estimate. Then find the quotient.

5. 75 ÷ 12.5

6. 544.6 ÷ 1.75

7. $0.78\overline{)0.234}$

MATHEMATICAL PRACTICE ⑥ Attend to Precision Algebra Evaluate using the order of operations.

8. 36.4 + (9.2 − 4.9 ÷ 7)

9. 16 ÷ 2.5 − 3.2 × 0.043

10. 142 ÷ (42 − 6.5) × 3.9

11. **GO DEEPER** Marcus can buy 0.3 pound of sliced meat from a deli for $3.15. How much will 0.7 pound of sliced meat cost?

_____

12. **THINK SMARTER** The table shows the earnings and the number of hours worked for three employees. Complete the table by finding the missing values. Which employee earned the least per hour? Explain.

Employee	Total Earned (in dollars)	Number of Hours Worked	Earnings per Hour (in dollars)
1	$34.02		$9.72
2	$42.75	4.5	
3	$52.65		$9.75

_____

_____

Connect to Science

Real World

Amoebas

Amoebas are tiny one-celled organisms. Amoebas can range in size from 0.01 mm to 5 mm in length. You can study amoebas by using a microscope or by studying photographic enlargements of them.

Jacob has a photograph of an amoeba that has been enlarged 1,000 times. The length of the amoeba in the photo is 60 mm. What is the actual length of the amoeba?

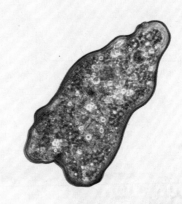

Divide 60 ÷ 1,000 by looking for a pattern.

60 ÷ 1 = 60

60 ÷ 10 = 6.0   The decimal point moves _____ place to the left.

60 ÷ 100 = _____   The decimal point moves _____ places to the left.

60 ÷ 1,000 = _____   The decimal point moves _____ places to the left.

So, the actual length of the amoeba is _____ mm.

13. **THINK SMARTER** Explain the pattern.

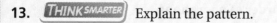

_____

14. **GO DEEPER** *Pelomyxa palustris* is an amoeba with a length of 4.9 mm. *Amoeba proteus* has a length of 0.7 mm. How many *Amoeba proteus* would you have to line up to equal the length of three *Pelomyxa palustris*? Explain.

_____

_____

Divide with Decimals

COMMON CORE STANDARD—6.NS.B.3
Compute fluently with multi-digit numbers and find common factors and multiples.

Estimate. Then find the quotient.

1. $43.18 \div 3.4$

$44 \div 4 = 11$

$$\begin{array}{r} 12.7 \\ 34\overline{)431.8} \\ -34 \\ \hline 91 \\ -68 \\ \hline 238 \\ -238 \\ \hline 0 \end{array}$$

2. $4.185 \div 0.93$

_____

3. $6.3\overline{)25.83}$

_____

4. $0.143 \div 0.55$

_____

Evaluate using the order of operations.

5. $4.92 \div (0.8 - 0.12 \div 0.3)$

_____

6. $0.86 \div 5 - 0.3 \times 0.5$

_____

7. $17.28 \div (1.32 - 0.24) \times 0.6$

_____

Problem Solving *Real World*

8. If Amanda walks at an average speed of 2.72 miles per hour, how long will it take her to walk 6.8 miles?

_____

9. Chad cycled 62.3 miles in 3.5 hours. If he cycled at a constant speed, how far did he cycle in 1 hour?

_____

10. **WRITE** *Math* Explain how dividing by a decimal is different from dividing by a whole number and how it is similar.

_____

_____

_____

Lesson Check (6.NS.B.3)

1. Elliot drove 202.8 miles and used 6.5 gallons of gasoline. How many miles did he travel per gallon of gasoline?

2. A package of crackers weighing 8.2 ounces costs $2.87. What is the cost per ounce of crackers?

Spiral Review (5.NBT.B.5, 5.NBT.B.7, 5.NF.B.3)

3. Four bags of pretzels were divided equally among 5 people. How much of a bag did each person get?

4. A zebra ran at a speed of 20 feet per second. What operation should you use to find the distance the zebra ran in 10 seconds?

5. Nira has $13.50. She receives a paycheck for $55. She spends $29.40. How much money does she have now?

6. A piece of cardboard is 24 centimeters long and 15 centimeters wide. What is its area?

© Houghton Mifflin Harcourt Publishing Company

FOR MORE PRACTICE GO TO THE Personal Math Trainer

Chapter 1 Review/Test

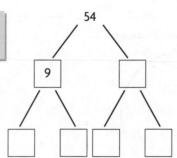 **Personal Math Trainer**
Online Assessment and Intervention

1. Use the numbers to complete the factor tree. You may use a number more than once.

2	3	6	9	27

Write the prime factorization of 54.

_____

2. For numbers 2a–2d, choose Yes or No to indicate whether the LCM of the two numbers is 15.

2a. 5, 3   ○ Yes   ○ No

2b. 5, 10   ○ Yes   ○ No

2c. 5, 15   ○ Yes   ○ No

2d. 5, 20   ○ Yes   ○ No

3. Select two numbers that have 9 as their greatest common factor. Mark all that apply.

(A) 3, 9

(B) 3, 18

(C) 9, 18

(D) 9, 36

(E) 18, 27

 Assessment Options
Chapter Test

4. The prime factorization of each number is shown.

$$15 = 3 \times 5$$
$$18 = 2 \times 3 \times 3$$

Part A

Using the prime factorization, complete the Venn diagram.

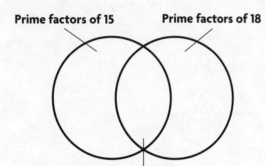

Prime factors of 15   Prime factors of 18

Common prime factors

Part B

Find the GCF of 15 and 18.

_____

5. For numbers 5a–5d, choose Yes or No to indicate whether each equation is correct.

5a. $222.2 \div 11 = 22.2$ ○ Yes   ○ No

5b. $400 \div 50 = 8$ ○ Yes   ○ No

5c. $1,440 \div 36 = 40$ ○ Yes   ○ No

5d. $7,236 \div 9 = 804$ ○ Yes   ○ No

Name _____

6. For numbers 6a–6d, select True or False for each equation.

6a. $1.7 + 4.03 = 6$ ○ True   ○ False

6b. $2.58 + 3.5 = 6.08$ ○ True   ○ False

6c. $3.21 - 0.98 = 2.23$ ○ True   ○ False

6d. $14 - 1.3 = 0.01$ ○ True   ○ False

Personal Math Trainer

7. **THINK SMARTER +** Four friends went shopping at a music store. The table shows the number of CDs each friend bought and the total cost. Complete the table to show the average cost of the CDs each friend bought.

Friend	Number of CDs Purchased	Total Cost (in dollars)	Average Cost (in dollars)
Lana	4	$36.68	
Troy	5	$40.55	
Juanita	5	$47.15	
Alex	6	$54.42	

What is the average cost of all the CDs that the four friends bought? Show your work.

_____

_____

8. The table shows the earnings and the number of hours worked for five employees. Complete the table by finding the missing values.

Employee	Total Money Earned (in dollars)	Number of Hours Worked	Earnings per Hour (in dollars)
1	$23.75		$9.50
2	$28.38	3.3	
3	$38.50		$8.75
4	$55.00	5.5	
5	$60.00	2.5	

9. The distance around the outside of Cedar Park is 0.8 mile. Joanie ran 0.25 of the distance during her lunch break. How far did she run? Show your work.

10. A one-celled organism measures 32 millimeters in length in a photograph. If the photo has been enlarged by a factor of 100, what is the actual length of the organism? Show your work.

11. GO DEEPER You can buy 5 T-shirts at Baxter's for the same price that you can buy 4 T-shirts at Bixby's. If one T-shirt costs $11.80 at Bixby's, how much does one T-shirt cost at Baxter's? Use numbers and words to explain your answer.

Name _____

12. Crackers come in packages of 24. Cheese slices come in packages of 18. Andy wants one cheese slice for each cracker. Patrick made the statement shown.

> If Andy doesn't want any crackers or cheese slices left over, he needs to buy at least 432 of each.

Is Patrick's statement correct? Use numbers and words to explain why or why not. If Patrick's statement is incorrect, what should he do to correct it?

13. There are 16 sixth graders and 20 seventh graders in the Robotics Club. For the first project, the club sponsor wants to organize the club members into equal-size groups. Each group will have only sixth graders or only seventh graders.

Part A

How many students will be in each group if each group has the greatest possible number of club members? Show your work.

Part B

If each group has the greatest possible number of club members, how many groups of sixth graders and how many groups of seventh graders will there be? Use numbers and words to explain your answer.

14. The Hernandez family is going to the beach. They buy sun block for $9.99, 5 snacks for $1.89 each, and 3 beach toys for $1.49 each. Before they leave, they fill up the car with 13.1 gallons of gasoline at a cost of $3.70 per gallon.

Part A

Complete the table by calculating the total cost for each item.

Item	Calculation	Total Cost
Gasoline	13.1 × $3.70	
Snacks	5 × $1.89	
Beach toys	3 × $1.49	
Sun block	1 × $9.99	

Part B

What is the total cost for everything before tax? Show your work.

Part C

Mr. Hernandez calculates the total cost for everything before tax using this equation.

Total cost = 13.1 + 3.70 × 5 + 1.89 × 3 + 1.49 × 9.99

Do you agree with his equation? Use numbers and words to explain why or why not. If the equation is not correct, write a correct equation.